Who Am I?
Prairie Animals

Written by Read With You
Center for Language Research and Development

Published by Read With You Publishing

Designed by Read With You Center for Language Research and Development

ISBN-13: 978-1-944710-77-4
ISBN-10: 1-944710-77-9

Printed in the United States of America.

I use my thick tail as a blanket.

Who am I?

xoɟ

I can spray a stinky smell.

Who am I?

skunk

I have four wings.

Who am I?

dragonfly

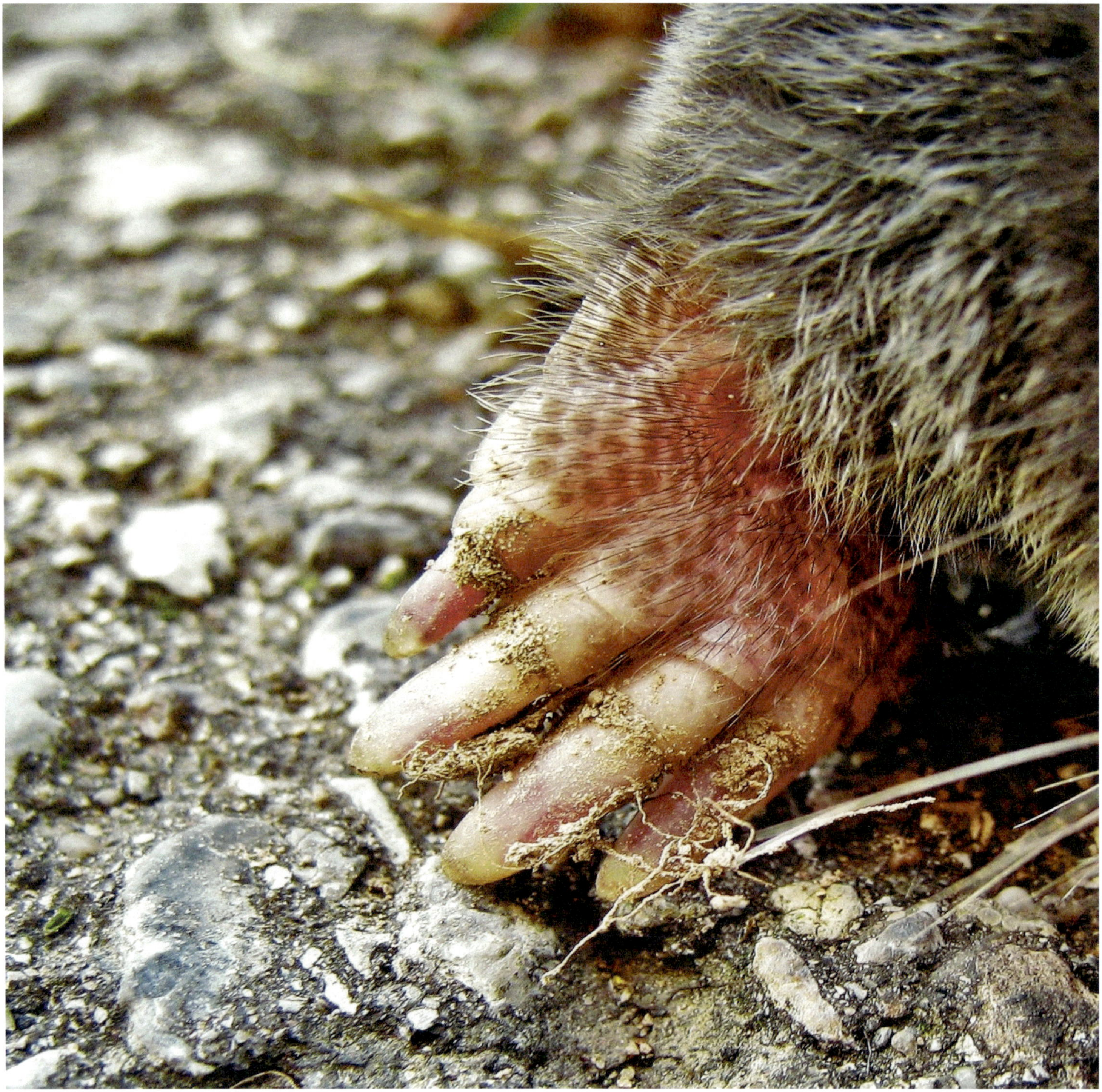

I dig holes and long tunnels.

Who am I?

mole

When scared, I sometimes play dead.

Who am I?

opossum

I’m huge, but I’m a fast runner.

Who am I?

bison

My long, sharp claws are made for digging.

Who am I?

badger

I can bark like dogs.

Who am I?

prairie dog

Fox

Skunk

Dragonfly

Mole

I am a fox.

I am a skunk.

I am a dragonfly.

I am a mole.

Opossum

Bison

Badger

Prairie Dog

I am an opossum.

I am a bison.

I am a badger.

I am a prairie dog.

We are prairie animals.

A prairie is a large area of flat, grassy land in North America.

Made in the USA
Monee, IL
26 April 2026

48945626R00017